Y0-BBE-861

ANIMALS THAT SURVIVE

THINGS ANIMALS DO

Kyle Carter

The Rourke Book Co., Inc.
Vero Beach, Florida 32964

© 1995 The Rourke Book Co., Inc.

All rights reserved. No part of this book may be reproduced or utilized in any form or by any means, electronic or mechanical including photocopying, recording or by any information storage and retrieval system without permission in writing from the publisher.

Edited by Sandra A. Robinson and Pamela J.P. Schroeder

PHOTO CREDITS
All photos © Kyle Carter

Library of Congress Cataloging-in-Publication Data

Carter, Kyle, 1949-
   Animals that survive / by Kyle Carter.
     p. cm. — (Things animals do)
   Includes index.
   ISBN 1-55916-114-0
   1. Animal defenses—Juvenile literature. [1. Animal defenses.]
I. Title. II. Series: Carter, Kyle, 1949-   Things animals do.
QL759.C37   1995
591.57—dc20   94-46847
                                                    CIP
                                                      AC

Printed in the USA

TABLE OF CONTENTS

Surviving	5
How Animals Survive	6
Hibernation	9
Camouflage	11
Pretending	14
Mimicking	16
Bright Means Bad	19
Suits of Armor	20
Fighting Back	22
Glossary	23
Index	24

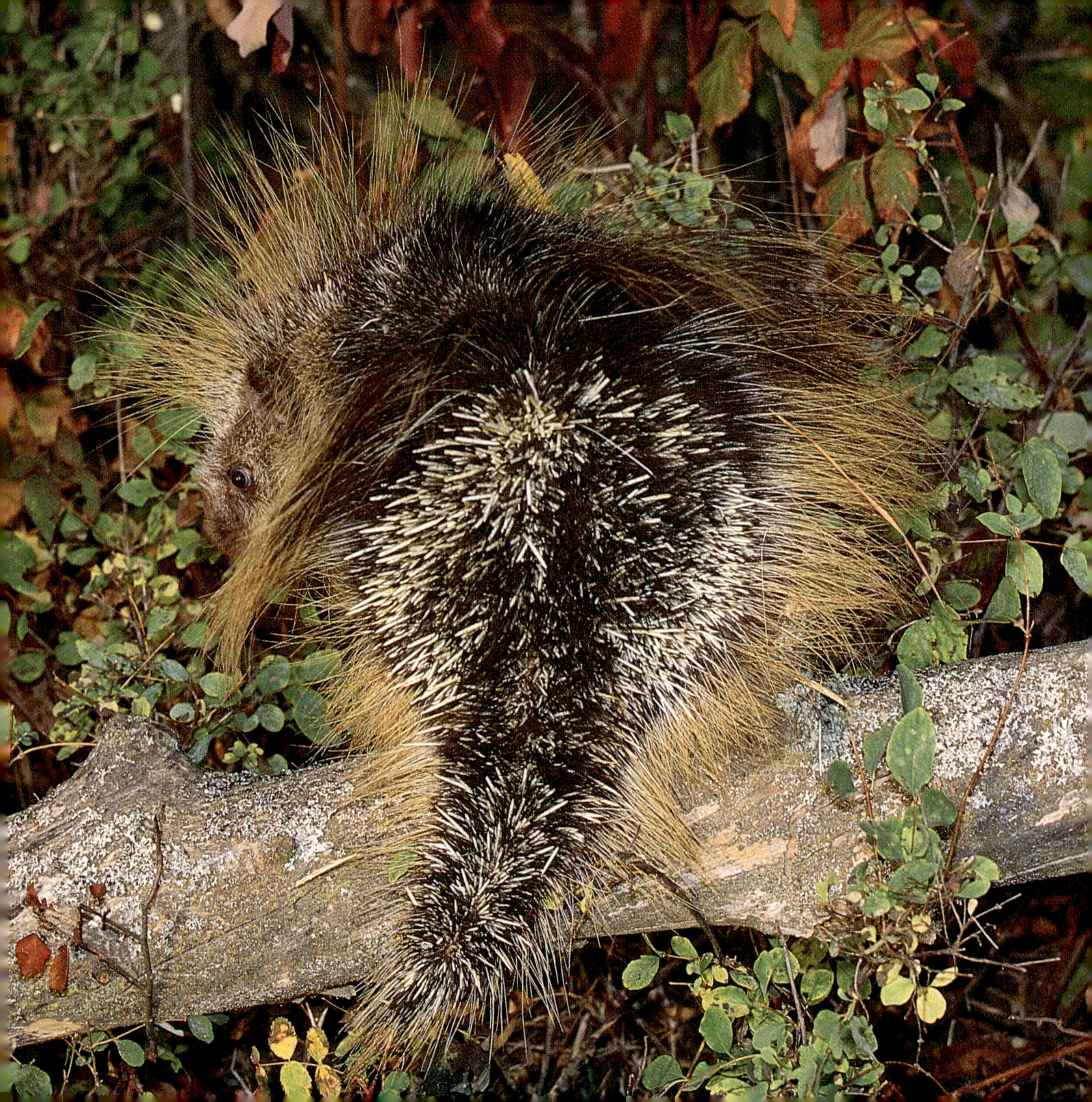

SURVIVING

Surviving—staying alive—is a full-time job for wild animals. Finding food is a large part of the battle for survival. However, for many animals survival depends not just on eating—it depends on not being eaten!

To avoid becoming a meal, animals have many unusual "tricks." They also have ways to survive when bad weather strikes.

The porcupine's tail is a nest of barbed quills that keep most predators away

HOW ANIMALS SURVIVE

Each animal has special tools to help it survive in its **habitat.** A habitat is the kind of place where an animal lives, such as a desert or pond. Each habitat has its own set of problems.

For example, seals live in icy cold ocean habitats. These animals survive thanks to a thick layer of blubber, or fat, on their bodies.

Other animals stay alive during dangerous times by traveling, **hibernating, camouflaging** themselves or **mimicking** another animal.

Tundra swans dodge the hardships of northern winters by traveling south where food is plentiful

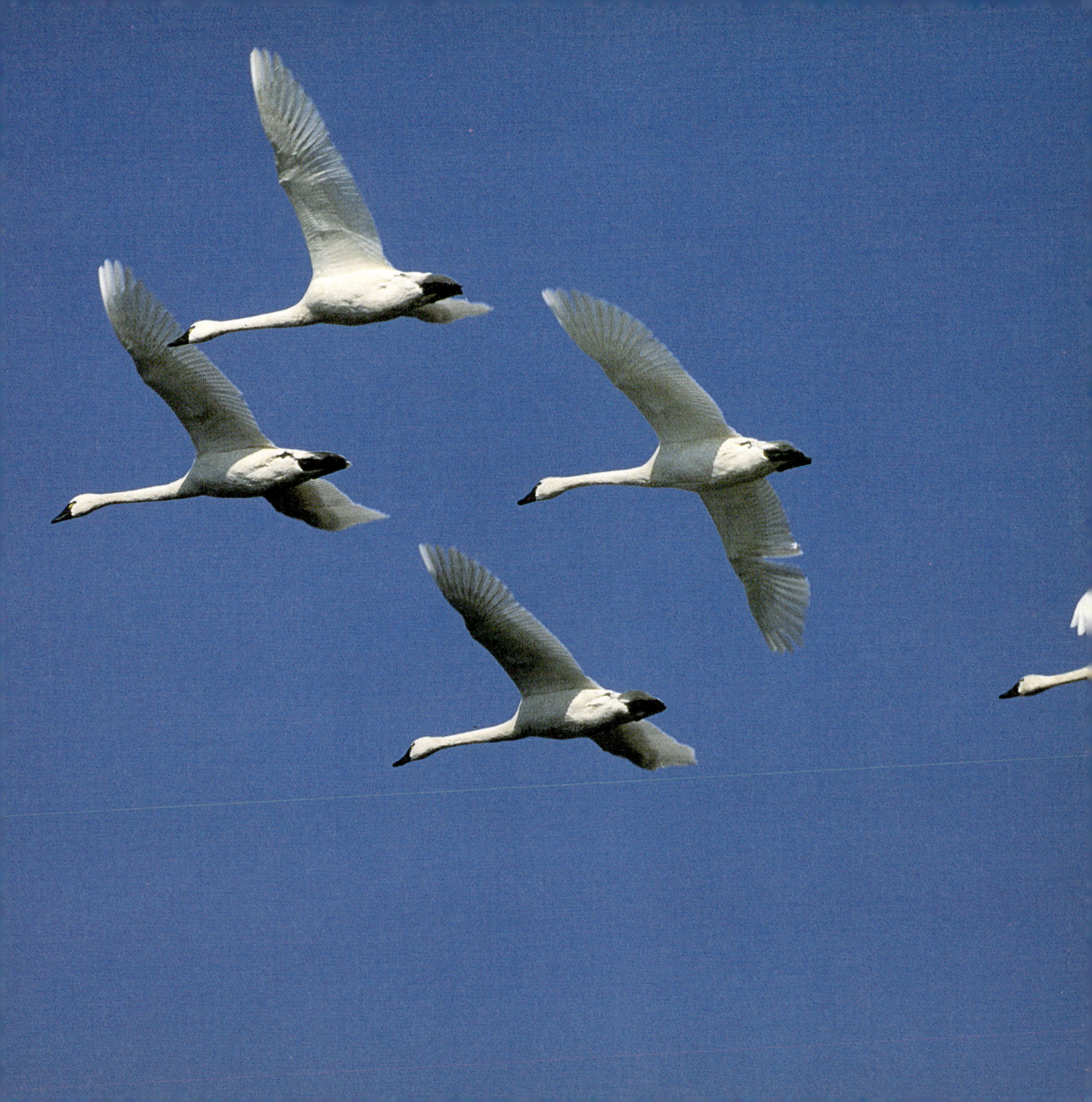

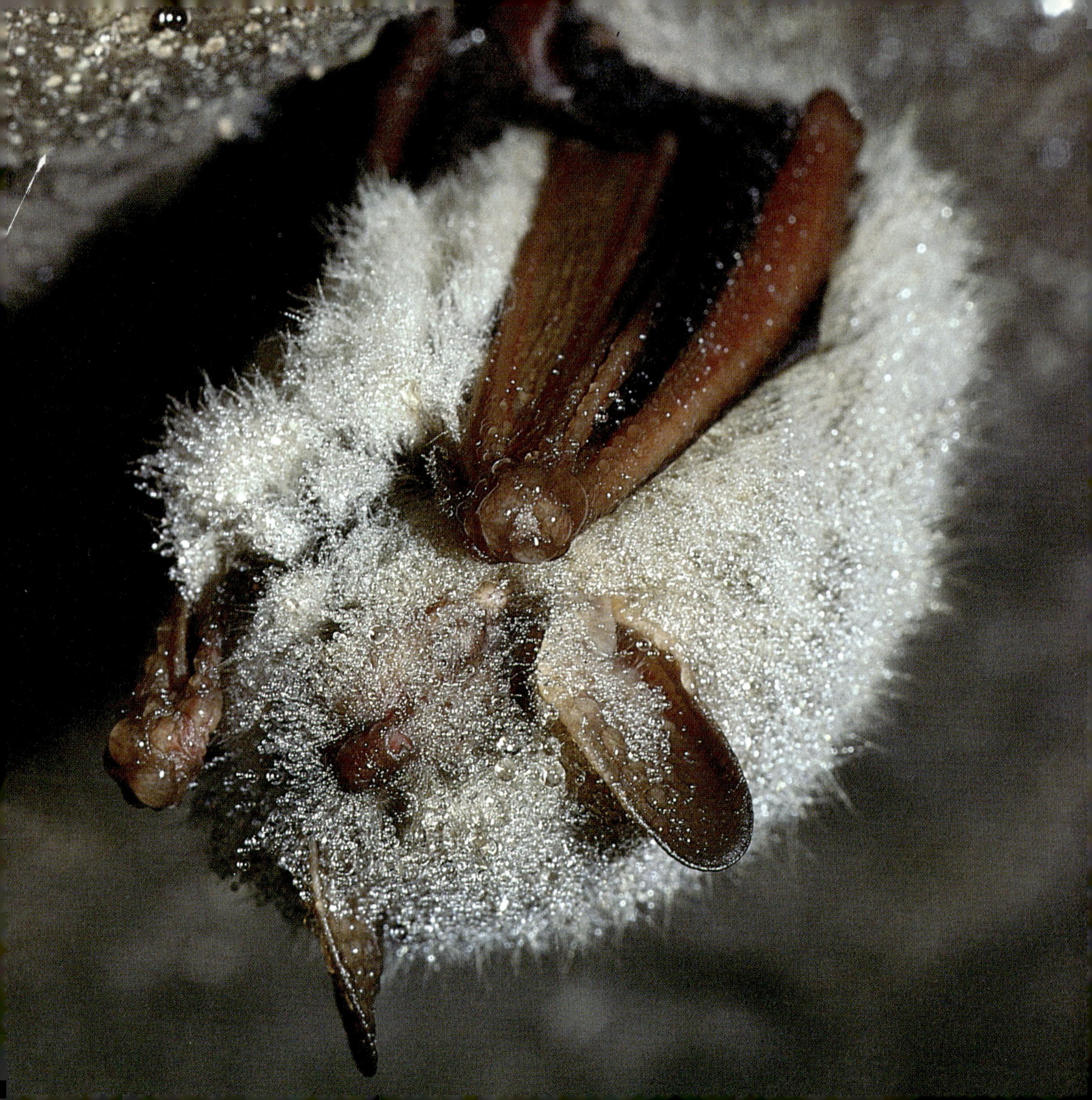

HIBERNATION

Many animals would not survive winter without a special "plan" for cold weather. When winter arrives, most animals either stay and hibernate or travel far away.

Hibernation is a deep sleep. An animal beds down in its cave or burrow in the fall and does not usually come out until spring.

Certain bears, ground squirrels, bats, snakes and turtles are among the hibernators. They live on body fat during their sleep. Their breathing and heartbeat slow down so the hibernators don't use much energy or food.

Water droplets glisten on a hibernating bat, snug in an Illinois cave

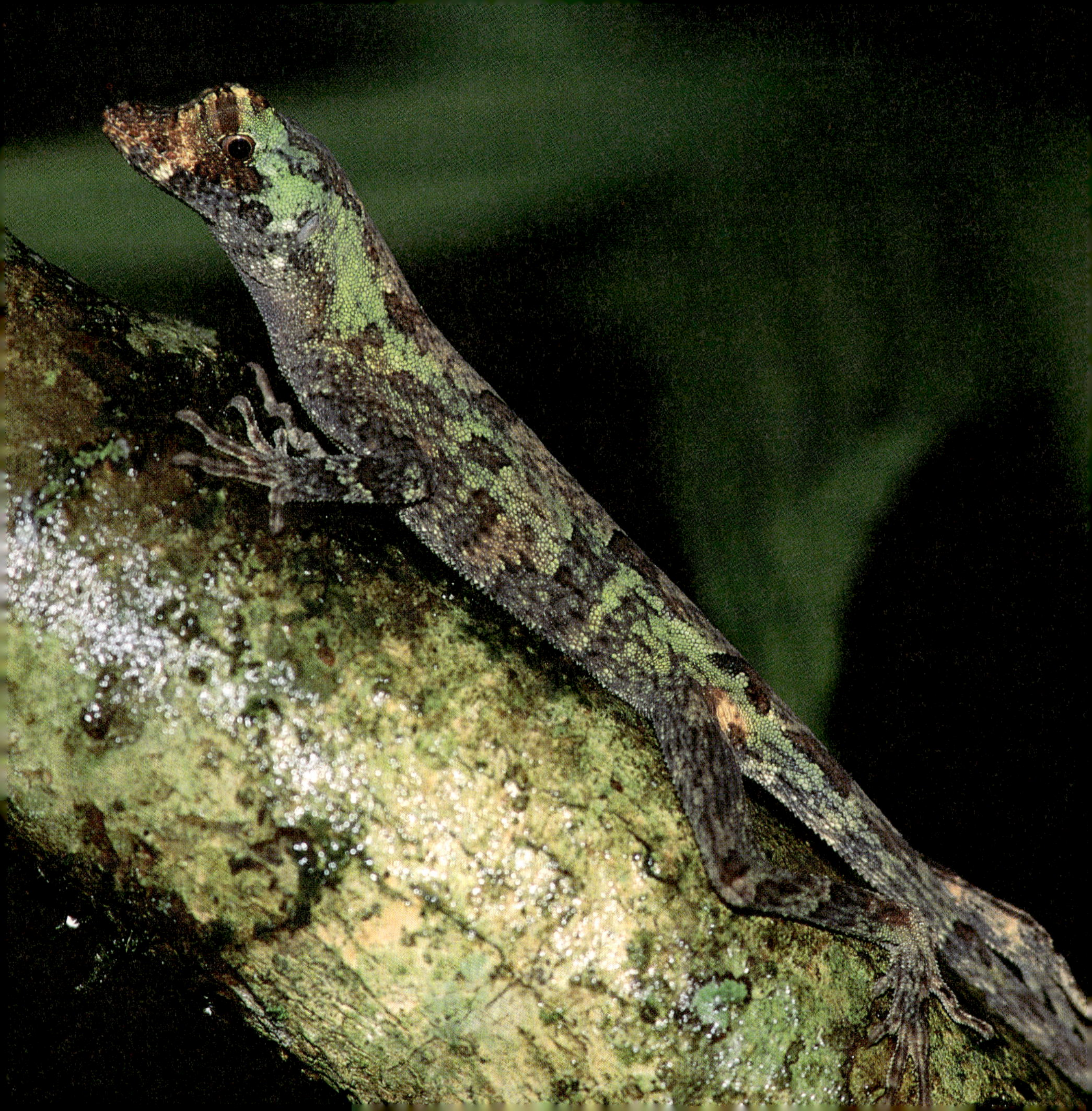

CAMOUFLAGE

Hunters wear camouflage clothing so they blend in with their surroundings. Some animals wear camouflage "clothing" every day. It helps them hide from **predators**—and avoid being eaten.

Certain insects look like leaves or twigs. The coloring of the scales of some lizards is a near-perfect match for tree bark or stones.

Snowshoe hares and weasels wear camouflage brown in summer. Each winter their fur coats become snow white!

When this Costa Rican anole hugs the branch, its coloring is a perfect match

Sea lions and other animals that live in herds and flocks warn each other of danger

Stick insects mimic the twigs of the forest where they live

PRETENDING

Another way to avoid being a meal is to pretend. A young horned owl, for example, fluffs its feathers to make itself look bigger and stronger than it is.

Hognose snakes are harmless. However, they can look and sound scary by puffing their heads and vibrating their tails on leaves.

Cranes, plovers and several other birds are pretenders of another kind. They'll fake a broken wing to lead predators away from their nests.

A killdeer, faking injury, helps her eggs survive by leading predators away from her nest

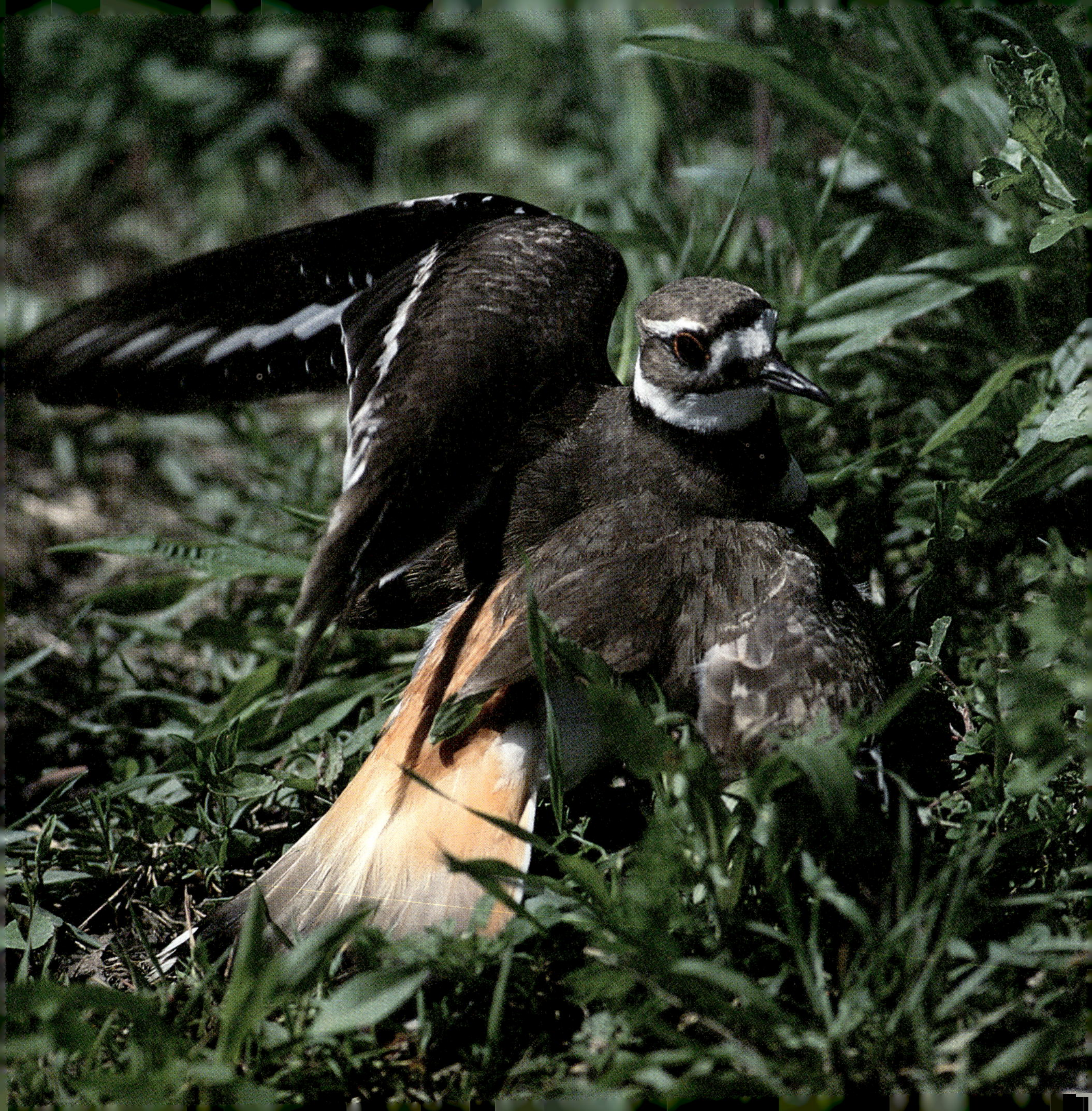

MIMICKING

A mimic pretends to be something it is not. Scientists think some animals mimic how others look to help them survive.

Birds usually don't eat more than one monarch butterfly because the monarchs taste bitter. The viceroy butterfly is a monarch look-alike. Scientists think that perhaps the viceroy mimics the monarch to avoid being a bird's lunch.

Many harmless snakes look like the deadly coral snake. Mimicking the colorful coral snake may help them scare predators.

The harmless scarlet kingsnake may scare some predators away because it looks so much like the deadly coral snake

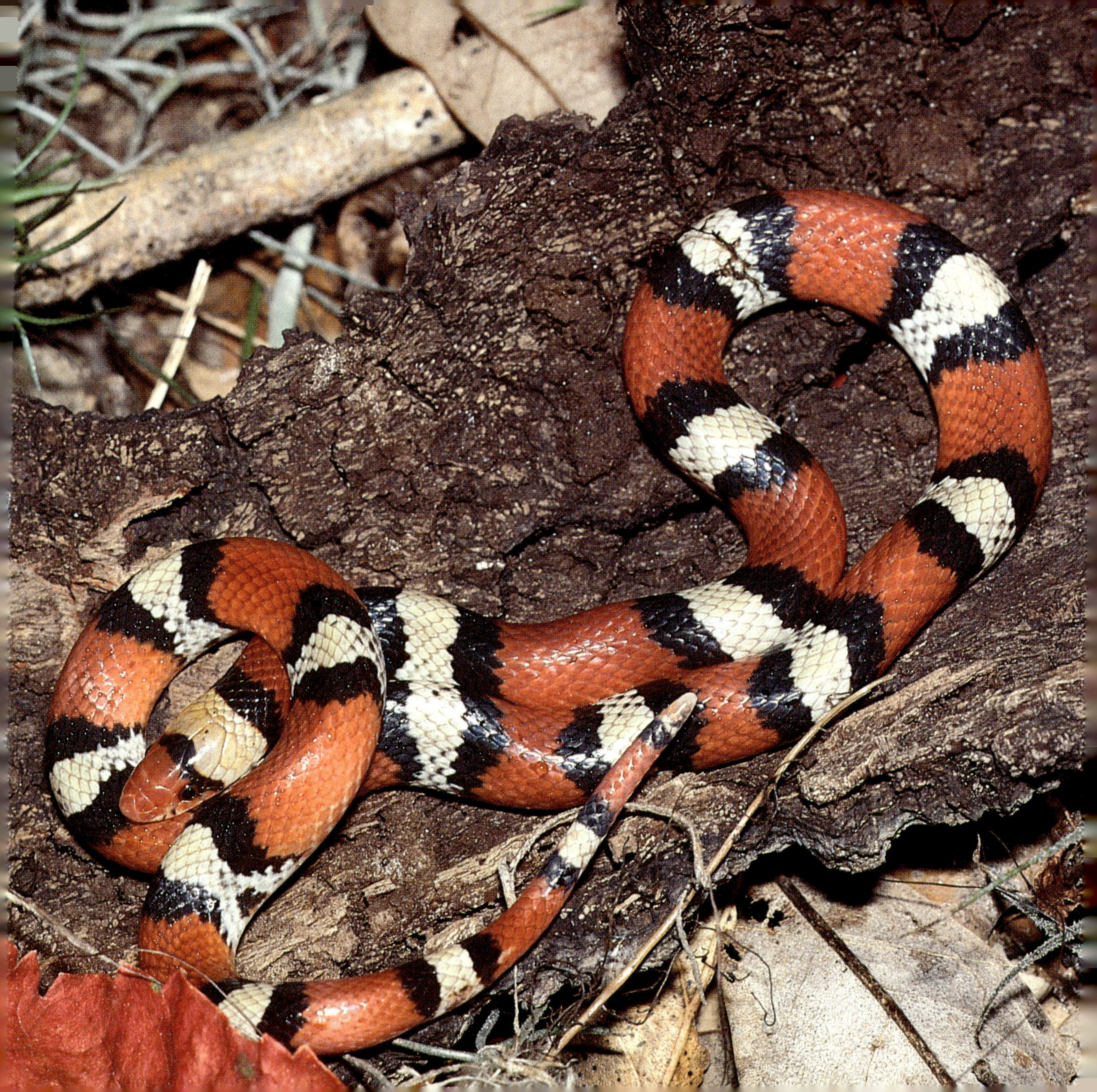

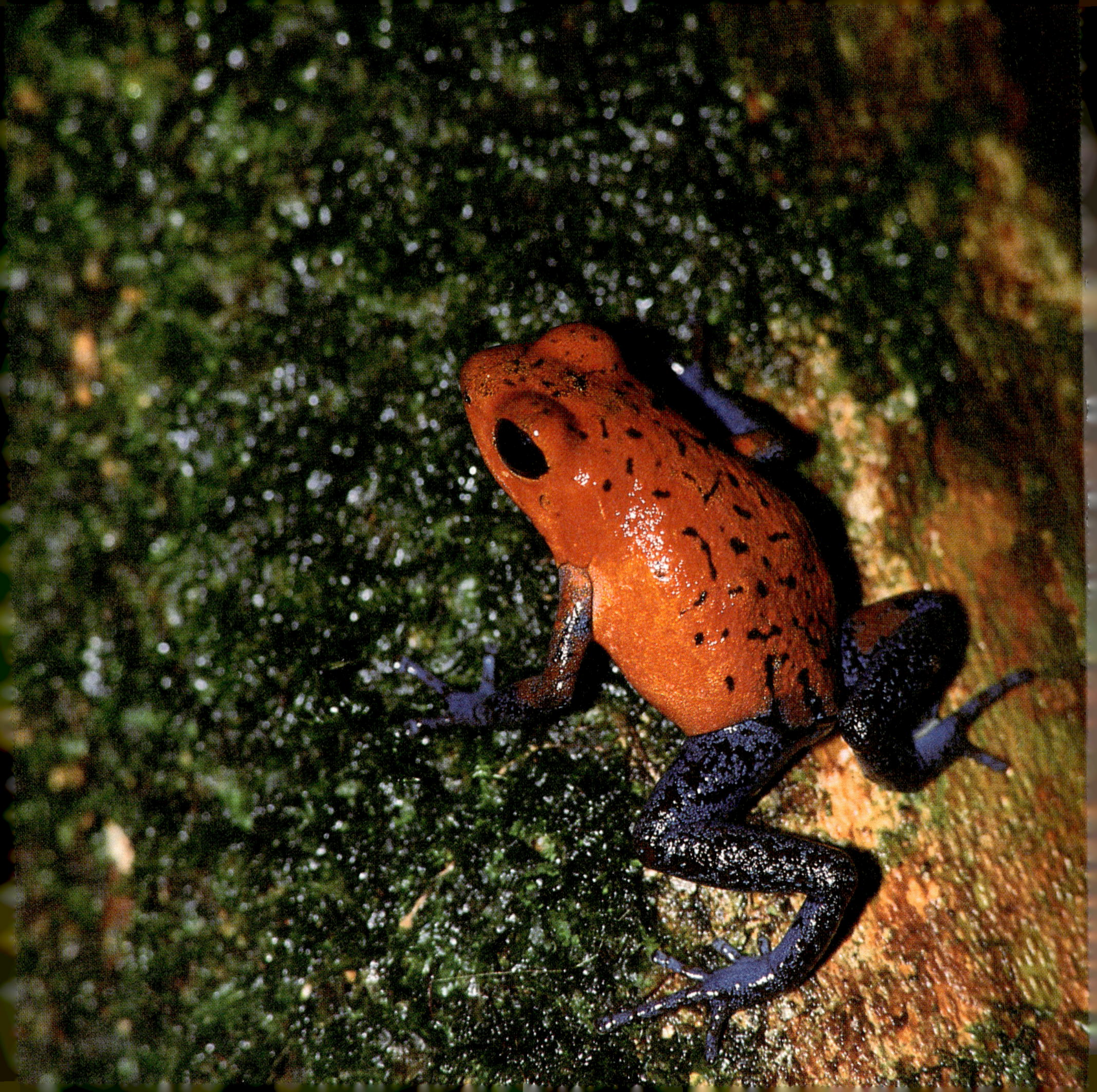

BRIGHT MEANS BAD

A few animals try to avoid trouble by standing out from the crowd. Instead of camouflage, they wear bright, bold colors.

To predators that can tell one color from another, bright may mean bad. The poison-arrow frogs of Central and South America, for example, are beautifully bright. They are also deadly. Natives use the frogs to poison the tips of hunting arrows.

The bright, bold black-and-white fur of the skunk tells predators to beware, too.

The poison-arrow frog's bold colors warn predators to beware

SUITS OF ARMOR

Nature helps certain slow-moving animals to survive with suits of armor. Animal armor isn't steel like the armor knights wore, but it does protect the wearer.

Turtles and snails grow shells as armor. Hermit crabs borrow empty snail shells and wear them. Armadillos are wrapped in a strong, leatherlike covering.

Armor doesn't guarantee survival, but it sure helps.

Armadillos are slow, but their leathery shell keeps them safe from many predators

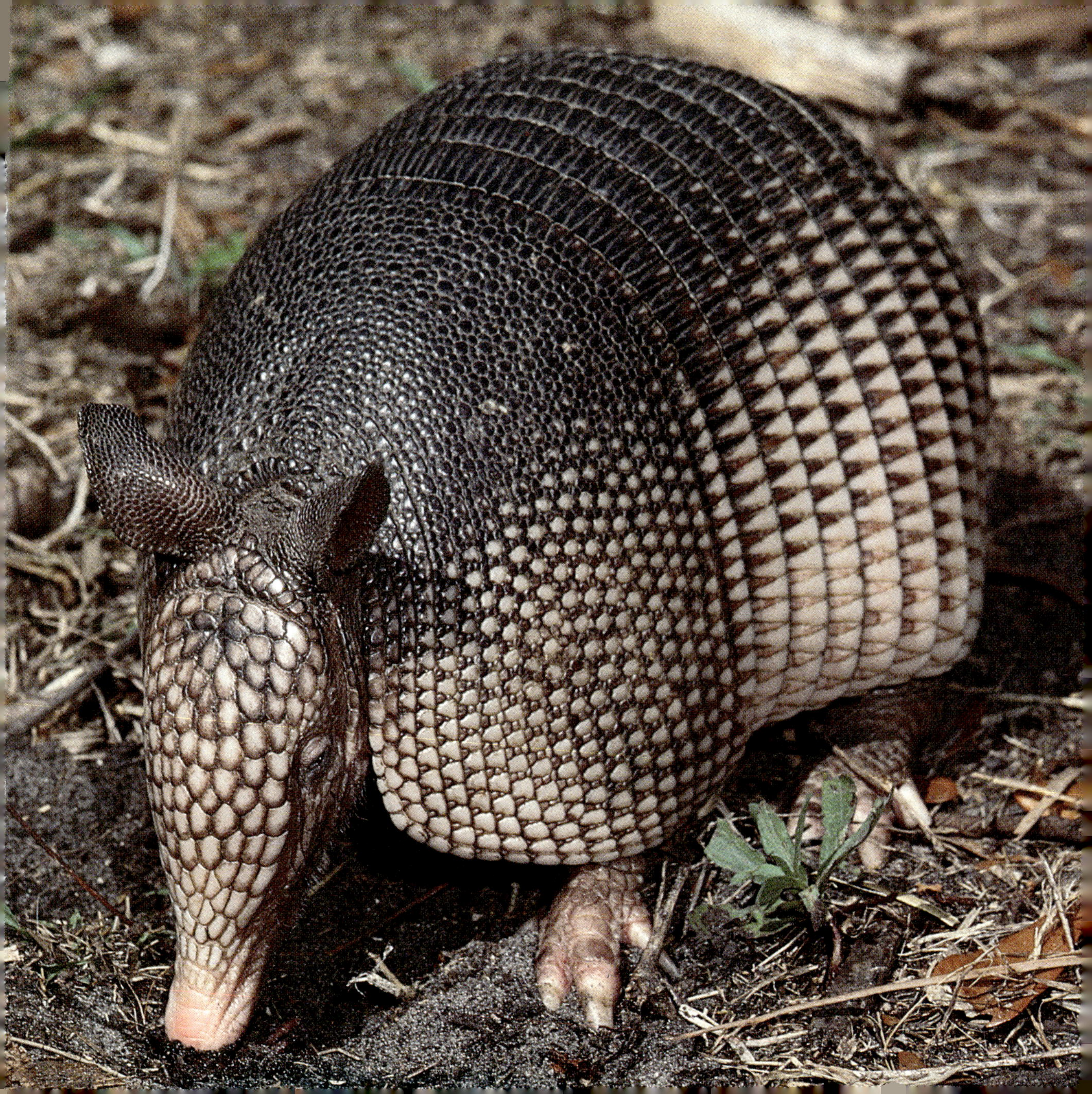

FIGHTING BACK

Most animals have some way to avoid being eaten. A rabbit runs. A squirrel climbs.

A porcupine can't run, and it's a slow climber. However, a porcupine's pincushion tail is loaded with sharp **quills.** Most porkies can save themselves by swatting a predator with the needlelike quills.

Sea urchins and many caterpillars are pincushions, too. Their spines often save them from being eaten by predators.

The sea hare uses yet another plan for survival. When threatened, it floods the water around it with bitter purple "ink." The predator turns away.

Glossary

camouflage (KAM o flahj) — coloring that allows an animal to blend into its surroundings

habitat (HAB uh tat) — the kind of place where an animal lives, such as a *desert*

hibernate (HI ber nate) — to enter a deep, sleeplike state for the winter

mimic (MIH mihk) — to copy the actions or appearance of another

predator (PRED uh tor) — an animal that kills other animals for food

quill (KWILL) — a hard, hollow, sharp hair

INDEX

armadillos   20
birds   14, 16, 22
butterfly
   monarch   16
   viceroy   16
cranes   14
food   5, 9
frogs, poison-arrow   19
hare, snowshoe   11
hibernation   9
hibernators   9
insects   11
lizards   11
mimic   16
mimicking   6
owl, horned   14

plovers   14
porcupine   22
predators   14, 16, 19, 20
quills   22
sea hare   22
seals   6
sea urchin   22
skunk   19
snails   20
snakes   9, 16
   coral   16
   hognose   14
turtles   9, 20
weasels   11